AF488459

Quark, Lepton, W and Z Masses
of Cosmos Theory and The Standard Model

Masses as Powers of 2 and π
Connection to Coupling Constants
Unification of Coupling Constants and Masses

Stephen Blaha Ph. D.
Blaha Research

Pingree-Hill Publishing
MMXXIV

To Margaret

Some Other Books by Stephen Blaha

SuperCivilizations: Civilizations as Superorganisms (McMann-Fisher Publishing, Auburn, NH, 2010)

All the Universe! Faster Than Light Tachyon Quark Starships & Particle Accelerators with the LHC as a Prototype Starship Drive Scientific Edition (Pingree-Hill Publishing, Auburn, NH, 2011).

Unification of God Theory and Unified SuperStandard Model THIRD EDITION (Pingree Hill Publishing, Auburn, NH, 2018).

The Exact QED Calculation of the Fine Structure Constant Implies ALL 4D Universes have the Same Physics/Life Prospects (Pingree Hill Publishing, Auburn, NH, 2019).

Passing Through Nature to Eternity ProtoCosmos, HyperCosmos, Unified SuperStandard Theory (Pingree Hill Publishing, Auburn, NH, 2022).

HyperCosmos Fractionation and Fundamental Reference Frame Based Unification: Particle Inner Space Basis of Parton and Dual Resonance Models (Pingree Hill Publishing, Auburn, NH, 2022).

The Cosmic Panorama: ProtoCosmos, HyperCosmos, Unified SuperStandard Theory (UST) Derivation (Pingree Hill Publishing, Auburn, NH, 2022).
Ultimate Origin: ProtoCosmos and HyperCosmos (Pingree Hill Publishing, Auburn, NH, 2022).

God and and Cosmos Theory (Pingree Hill Publishing, Auburn, NH, 2023).

Newton's Apple is Now The Fermion (Pingree Hill Publishing, Auburn, NH, 2023).

Cosmos Theory: The Sub-Particle Gambol Model (Pingree Hill Publishing, Auburn, NH, 2023).

Cosmos-Universe-Particle-Gambol Theory (Pingree Hill Publishing, Auburn, NH, 2024).

Fractal Cosmos Curve: Tensor-based Cosmos Theory (Pingree Hill Publishing, Auburn, NH, 2024).

The Eternal Form of Cosmos Theory Third Edition (Pingree Hill Publishing, Auburn, NH, 2024).

Fundamental Constants of Cosmos Theory and The Standard Model (Pingree Hill Publishing, Auburn, NH, 2024).

Available on Amazon.com, bn.com Amazon.co.uk and other international web sites as well as at better bookstores.

CONTENTS

FIGURES and TABLES

Introduction

Cosmos Theory has evolved from a specification of spaces and their associated arrays of dimensions to the consideration of the consistency condition that limits the set of Physical spaces to ten. Physical spaces support universes including ours. Our 4 dimension universe has sets of fermions, vector bosons, scalar bosons, and so on that are defined using Internal Symmetry groups. These symmetry groups arise as amalgams of sets of eight fundamental representation dimensions.

The proof of Cosmos Theory at the multi-space, multi-universe levels would be difficult to say the least. And likely to be so indefinitely into the future – although there are some hints of phenomena beyond our universe.

Recently we developed a consistency condition that limits the number of Physical spaces to ten. Remarkably, this derivation led directly to the forms of physical constants such as the Fine Structure Constant, and the coupling constants of the ElectroWeak and Strong Interactions. The coupling constants were shown to consist of the base, e, of the natural logarithms times factors consisting of powers of two. We were able to show that these factors of two directly originated in the number of degrees of freedom of fermions. The values of coupling constants are now understood. The closeness of their expressions to features of the Limos spaces of Cosmos Theory can only be interpreted as direct experimental support for Cosmos Theory within our universe.

This book considers other major global constants of our universe, namely, elementary particle masses. Working from experimentally known eight fundamental fermion (current) masses, the book exhibits mass patterns exist that lead to accurate estimates of fermion masses using powers of two and π. The forms of these estimates resemble the forms of coupling constants. These mass value estimates give significant additional circumstantial evidence for Cosmos Theory within our universe.

An interesting point that arises: Coupling Constants have a factor of e, masses have a factor of π to a power, and the consistency condition unites e and π in $e\pi/4$ which plays the role of a dimension in the consistency condition. Note that $e\pi/4 \cong 2$ with 2 being the power sequence variable in dimension specifications in Cosmos spaces. One might say the dimensions may figuratively be viewed as the result of the "union" of coupling constants and masses.

The list of evidence for Cosmos Theory is further augmented by the calculation of the Weinberg angle and the vector boson Z and W masses within the Cosmos Theory framework based on Coupling Constant expressions and fermion mass expressions.

The book also makes projections of fermion masses to higher generations.

The mass calculations presented here greatly strengthen the experimental support for Cosmos Theory. Cosmos thought is being brought down to earth!

1. Current Masses in Cosmos Theory and The Standard Model

In Blaha (2024f) we developed a theory of coupling constants that yielded the known coupling constants of The Standard Model together with extremely accurate mathematical expressions for each of the known interactions. These expressions were shown to have a basic form within a Cosmos Theory perspective as a constant e (the base of natural logarithms) times the number of degrees of freedom (spins) for the fermions of the fundamental representation of each coupling constant's internal symmetry group. Thus they have a fundamental basis.

We now turn to the other known set of constants in The Standard Model and Cosmos Theory – the eight fermion (quark and lepton) masses of the first generation of fermions. We will show that fairly simple patterns exist in these fermion masses that lend themselves to being expressed as simple mathematical formulas. The form of these formulas partly reflects the form of previously derived coupling constants' values.

The fermions of the fermion mass spectrum, so derived, express the masses in the form of integer mathematical factors times powers of π. In a sense the appearance of π in masses complements the appearance of e in coupling constant values. Note that $e\pi/4$ is approximately 2, which is the power sequence variable in the dimension array sizes (2^{r+4}) of Cosmos Theory spaces. Thus masses and Cosmos Theory dimensions have similar expressions.

We begin with the coupling constants of Fig. 1.1 found in Blaha (2024f).

		E S T I M A T E		**E X P E R I M E N T**		
Interaction	**Expression**	$g^2/4\pi$ **Value**	**g**	**Known**[1] **Value** $g^2/4\pi$		**Deviation**
U(0) α_0	$e^2/4096$	0.0018036	-	-		-
U(1) $\alpha_1 = \alpha$	$e^2/1024$	0.00721438	0.303	0.0072973525643		1.15%
SU(2) $\alpha_2 = g^2/4\pi$	$e^2/256$	0.0289	0.63	0.0316		9.3%
SU(3) $\alpha_3 = \alpha_S$	$e^2/64$	0.115	1.21	0.117		1.7%
SU(4)^2 $\alpha_4?$	$e^2/16$	0.462	2.4?	0.458		0.087%

Figure 1.1. The Coupling Constants for first generation internal symmetries in Cosmos Theory and The Standard Model.

Note the interrelationships embodied in the coupling constant expressions such as:

$$\alpha_4{}^2 - e^2\alpha_2 \tag{1.1}$$

---

[1] All coupling constant values are based on data from Particle Data Group 2022.
[2] This value is based on the "doubling trend" seen in the three known coupling constants above.

1.1 Fermion Generations

The Standard Model *defines* fermion generations (Fig. 1.2). Cosmos Theory and, in particular, The Unified SuperStandard Theory (UST) within it define generations in a manner based on the form of the r = 4 dimension array of the Cosmos Spectrum.[3] (Fig. 1.3) The UST has four layers, which each contain four generations.[4] Each generation has eight "normal" fundamental fermions. (There is also a corresponding eight Dark fundamental fermions in the Dark matter sector.) Thus there is a total of 4×4×8 = 128 normal fermions and 128 Dark fermions.

Standard Model Generations

Generation:	1	2	3
	u	c	t
	d	s	b
	e	μ	τ
	ν_e	ν_μ	ν_τ

Figure 1.2. The Standard Model generations.

We will consider the *Current* mass spectrum of the eight first generation "normal" fermions. We will show they exhibit interesting regularities, which support the UST (and Cosmos) view of generations and also provide a simple, surprising, numerical fermion spectrum of current masses.

---

[3] See Blaha (2023d) and (2018e) for details.

[4] The four layers support the author's Layer groups and the four generatioons per layer support the author's Generation groups. See Appendix 1-A for a description of these groups, which were originally introduced some time ago by the author.

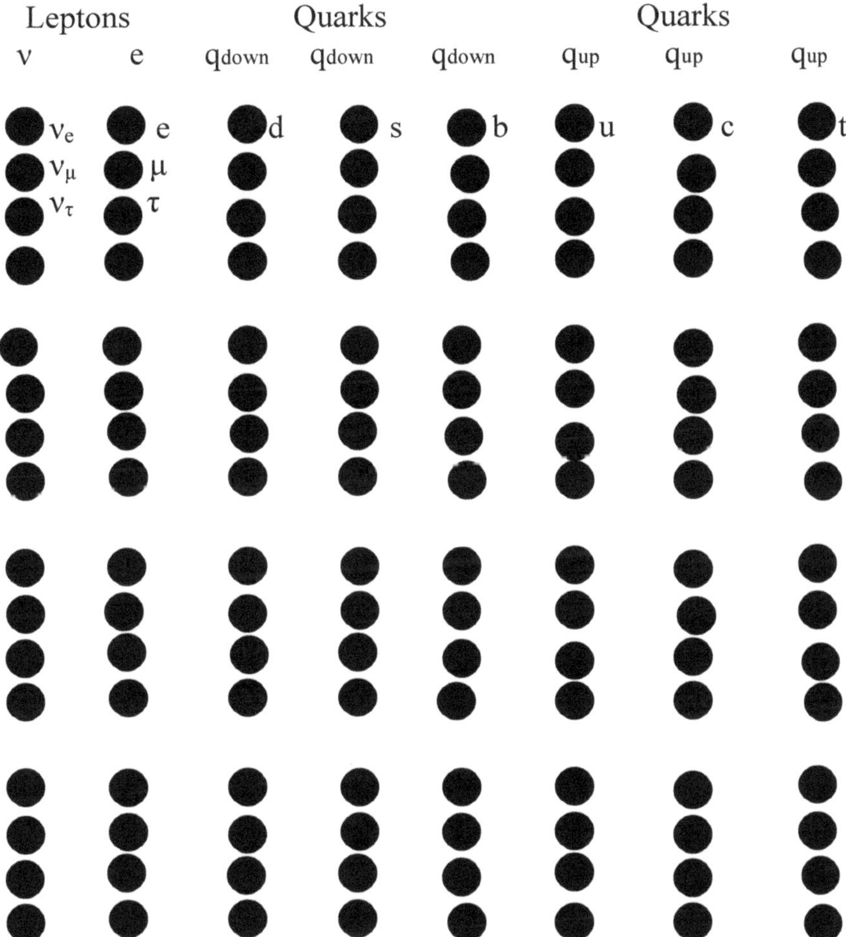

Figure 1.3. Four UST (and Cosmos) layers with each consisting of four generations of eight normal and eight Dark fermions. From Blaha (2023d) and the author's earlier books.

$\nu_e = 3.878 \times 10^{-11}$ GeV/c^2

Our ν_e mass estimate = .03878 ev/c^2 is consistent with known data estimates.

$\nu' = \nu_e \, (\alpha_0^2)^{-1} = 0.03878$ ev/c$^2 \times 4096^2/e^2 = 8.8 \times 10^{-5}$ GeV/c^2 where e = 2.718.

We scale ν_e by $(\alpha_0^2)^{-1}$. This scaling puts the neutrino mass on the same footing as the other masses. It may reflect the need to scale the neutrino mass by the U(0) "symmetry." Later we suggest a further need to scale mass values based on the ratio of SU(2) mass effects to SU(3) mass effects, which appear to vary as a function of fermion mass.

e $= 0.511 \times 10^{-3}$ GeV/c^2 = electron mass

u 1.76 mev/c2 $= 1.76 \times 10^{-3}$ GeV/c^2

We scale the u mass by a factor of 4/5 – again to put it on a similar footing as the other masses. Note: the u mass is 2.2 +0.5 -0.4 $\times 10^{-3}$ GeV/c^2 The 1.76 value is consistent with the lower bound 1.8.

c 1.27 GeV/c^2

t 172.76 GeV/c^2

d 3.76×10^{-3} GeV/c^2

We scale the d mass by a factor of 4/5 – again to put it on a similar footing as the other masses. Note: the d mass is 4.7 +.5 - .3 $\times 10^{-3}$ GeV/c^2 The 3.76 value is consistent with the lower bound 4.4.

s 95×10^{-3} GeV/c^2

b 4.18 GeV/c^2

Figure 1.4. Fundamental first generation current fermion masses. Each fundamental fermion symbol represents its mass.

Appendix 1-A. Generation and Layer Groups[5]

1-A.1 U(4) Generation Groups

In the Big Bang all particles were massless and all symmetries unbroken. Hence the four Normal particle number symmetries, and the four Dark particle number symmetries, are all "conserved" in the Big Bang. Afterwards conservation laws are then broken.

We define two particle number operators for normal up-quark particles and down-quark particles, B_{uq} and B_{dq}. Similarly we define two particle number operators for normal species "e" (electron) particles and species "v" particles, B_e and B_v. Similarly we define Dark matter equivalents:[6] B_{De}, B_{Dv}, B_{Duq}, and B_{Ddq}.

In the absence of symmetry breaking these fermion particle number operators would be conserved. Thus there are two sets of "diagonal" operators with associated U(4) groups for the Normal and Dark sectors. They are part of the Normal U(4) Generation Group and the Dark U(4) Generation Group.

The fermion fundamental representation of a U(4) group has four fermions. U(4) has rotations, and also interactions of the form $\overline{\Psi}\gamma\cdot B\cdot T\Psi$ where Ψ is a fermion four-vector, B is a 16 component U(4) gauge field, and T consists of 16 component 4×4 U(4) arrays.

In the case of the Generation Group the gauge fields have electric charge zero. Since the four species have different electric charges (1, 0, 2/3, -1/3) the U(4) gauge boson fields cannot mix the fermions of different species. Generation Group interactions are diagonal[7] in fermion species (e, v, up-quark, and down-quark species).

Consequently the U(4) Generation Group for each layer[8] must have a reducible representation D consisting of a set of four fundamental U(4) representations, D_e, D_v, D_{upq}, and D_{dnq}, appearing in blocks along the diagonal of D. Each block is a separate U(4) irreducible representation for a species (due to the electric charge superselection rule.) There is a U(4) Generation group for each of the four layers of the Normal and Dark sectors totaling to 8 generation groups.

There are four generations of each species in the Normal and in the Dark matter sectors. The four generations for each fermion species: e, v, up-quark, and down-quark each furnish a U(4) fundamental representation within the reducible representation D. The fourth generation of normal fermions has not as yet been found due to their extremely large masses.

---

[5] From Blaha (2019g), (2018e) and (2023d).

[6] By analogy, we assume that there are four species of Dark matter: charged Dark leptons, neutral Dark leptons, Dark up-type quarks, and Dark down-type quarks. Thus we are led to the Dark particle numbers: Dark Baryon Numbers, and Dark Lepton Numbers shown above.

[7] ElectroWeak interactions can cross between species due to their charged gauge vector bosons.

[8] For the Normal and Dark sectors separately.

The Generation Group rotates the fundamental fermions of each fundamental representation separately for each of the four species of each of the four layers.[9] Thus the Generation Group guarantees that all generations of each species have the same electric charge and other quantum numbers.

The U(4) Generation Group also specifies a gauge field interaction among the fermions of its fundamental representation, species by species, for both Normal and Dark sectors. The form of the interactions for the Normal sector for each fermion layer is:

$$g_e \overline{\Psi}_e \gamma \cdot B_e \cdot T\Psi_e + g_v \overline{\Psi}_v \gamma \cdot B_v \cdot T\Psi_v + g_{upq} \overline{\Psi}_{upq} \gamma \cdot B_{upq} \cdot T\Psi_{upq} + g_{dnq} \overline{\Psi}_{dnq} \gamma \cdot B_{dnq} \cdot T\Psi_{dnq}$$

$$(1\text{-}A.1)$$

where g_e … are coupling constants, the gauge vector fields are B_e … , and the Ψ_e … are 4-vectors of fermions of the four generations of each species in a layer.

The gauge vector bosons of the Generation Group have large masses. If the conservation of the fermion particle numbers is broken then we view it as a consequence of Generation Group symmetry breaking.

Generation Group rotations guarantee the internal quantum numbers of each generation of each species are the same since symmetry breakdown is not present at the instant of the Big Bang.

The above discussion applies similarly to the Dark sector. Thus there are 8 Generation Groups in total.

See Blaha (2019g) and (2018e) for a detailed discussion of the Generation Groups.

1-A.2 U(4) Layer Groups

The set[10] of particle number operators can be extended if we take account of the fourfold fermion generations.

We can subdivide the above particle number sets into four additional particle numbers *per generation*. For the ith generation (of the four generations) we define

L_{ie} – The "e" species particle number for the ith generation
L_{iv} – The v species particle number for the ith generation
L_{iuq} – The up-quark species particle number for the ith generation
L_{idq} – The down-quark species particle number for the ith generation

L_{iDe} – The Dark "e" species particle number for the ith generation
L_{iDv} – The Dark v species particle number for the ith generation
L_{iDuq} – The Dark up-quark species particle number for the ith generation
L_{iDdq} – Dark down-quark species particle number for the ith generation

---

[9] There are separate Generation groups for each layer.

[10] Here again, in the Big Bang all particles were massless and all symmetries unbroken. Hence particle numbers are "conserved" in the Big Bang. Conservation is then broken afterwards in most cases.

for each generation i = 1, 2, 3, 4. Individual fermions have positive $L_{ia} = +1$ values and antifermions have negative $L_{ia} = -1$ values for each species.

At this point we have a set of four particle number operators for each of four generations (i = 1, 2, 3, 4) of fermions in the Normal sector and similarly in the Dark sector. We then define a U(4) group framework for each set of particle numbers.

The only way to specify fundamental representations for each of the four sets in a sector is to assume there are four layers, with each layer having four generations, and with a fundamental U(4) representation defined for each generation composed of fermions from each layer. Thus there are four Layer Groups for each Normal and each Dark sector: a Layer Group for generation 1, a Layer Group for generation 2, and so on.

The Layer Groups are also "split" by species due to the electric charge superselection rule. Each Layer Group is diagonal in the four fermion species. All their gauge fields are electrically neutral. Thus the Layer Group fundamental representations of the Normal sector total 16.[11] The Layer Group fundamental representations of the Dark sector also total 16. There are 4 Normal Layer Groups and 4 Dark Layer Groups.

Consequently each of the four U(4) Layer Groups in the Normal fermion sector has a reducible U(4) representation D_j for j = 1, 2, 3, 4. Each reducible representation is composed of four irreducible U(4) representations for each species due to the electric charge superselection rule:

$$D_j = D_{je} + D_{jv} + D_{jupq} + D_{jdnq},$$

for j = 1, 2, 3, 4.

There are four layers of each species in the Normal and in the Dark matter sectors. The second, third and fourth layers of normal fermions has not as yet been found due to their extremely large masses.

A Layer Group rotates the fundamental fermions of each fundamental representation separately for each of the four species of each of the four generations.

The Layer Groups guarantee that all layers of each species have the same electric charge and other quantum numbers.

Each U(4) Layer Group also specifies a gauge field interaction among the fermions of its fundamental representation, species by species, for both Normal and Dark sectors. The form of the interactions is:

$$g_{ei}\overline{\Psi}_{ei}\gamma\cdot C_{ei}\cdot T\Psi_{ei} + g_{vi}\overline{\Psi}_{vi}\gamma\cdot C_{vi}\cdot T\Psi_{vi} + g_{upqi}\overline{\Psi}_{upqi}\gamma\cdot C_{upqi}\cdot T\Psi_{upqi} + g_{dnqi}\overline{\Psi}_{dnqi}\gamma\cdot C_{dnqi}\cdot T\Psi_{dnqi}$$

$$(1\text{-}A.2)$$

for i = generation = 1, ... , 4, where g_{ei} ... are coupling constants, the gauge fields are C_{ei} ... , and the Ψ_{ei} ... are 4-vectors of fermions formed of the ith generation fermions in each layer of each species.

---

[11] Four Layers groups irreducible representations for the generations × four species = 16 irreducible representations.

The gauge vector bosons of the Layer Groups also have large masses. If the conservation of the fermion particle numbers is broken then we view it as a consequence of Layer Groups symmetry breaking.

Layer Group rotations guarantee the internal quantum numbers of each layer of each species are the same since symmetry breakdown is not present at the instant of the Big Bang.

The above discussion applies similarly to the Dark sector. There are 16 Dark Layer Groups.

Fig. 7.3 shows the fundamental fermion spectrum with the representations of the Generation groups and Layer groups indicated.

Experimentally, we know of three generations of fermions—the lowest 3 generations of the lowest level. The remaining 4th generation and three layers of fermions are of much higher mass and are yet to be found.

See Blaha (2019g) and (2018e) for a detailed discussion of the Layer Groups. We note in passing that the symmetries of these number operators are badly broken. Yet the underlying group structure remains.

2. Analysis of the First Generation Fermion Current Masses

At first glance the first generation mass spectrum of Fig. 1.4 does not appear to exhibit any regularities. In this chapter we will develop a new view of the spectrum based on the diagram in Fig. 2.1.

This diagram portrays the fermion masses on two levels based on the ElectroWeak separation of fermions in pairs. On each level the masses are ordered in increasing value. The diagram inverts the u and d level locations based on the overall ordering of top level masses as above the corresponding low level masses. Each ElectroWeak pair has the more massive above the less massive.

We introduce vertical arrows pointing to the lower mass of corresponding pairs. We introduce "angled" arrows between consecutive pairs pointing to the larger mass of adjacent pairs. This procedure generates a path between masses that leads to surprising regularities in fermion mass relations.

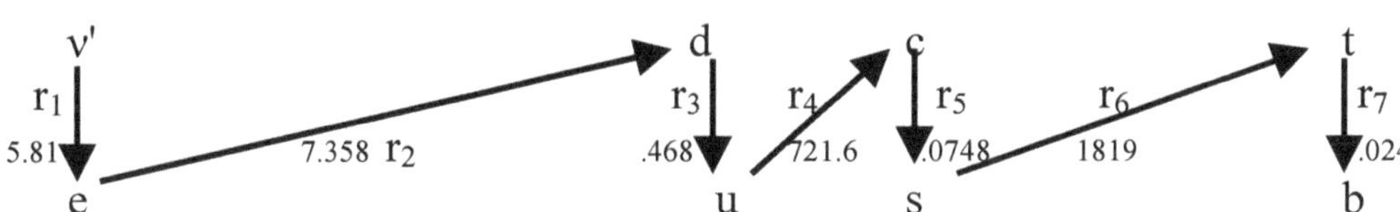

Figure 2.1. Fundamental first generation current fermion masses. Each fundamental fermion symbol represents its mass.

Fig. 2.1 displays ratios denoted r_i where i =1, 2, … , 7 that we will now define and use to relate ratios for the purpose of generating relationships between masses. We define the ratios with their corresponding numeric values obtained directly from Fig. 1.4.

$$r_1 = e/v' = .131/.02254 = 5.81 \tag{2.1}$$
$$r_2 = d/e = 7.358$$
$$r_3 = u/d = 0.468$$
$$r_4 = c/u = 721.6$$
$$r_5 = s/c = 74.8\times10^{-3} = 0.0748$$
$$r_6 = t/s = 1.819\times10^{3} = 1819$$
$$r_7 = b/t = 0.024$$

We can use these ratios and values to define fermion masses. Note first

$$v' = v_e \, (\alpha_0^{\,2})^{-1}$$
$$r_2/r_1 = 1.27 \cong 4/\pi \tag{2.2}$$

and then we use eq. 2.1 to obtain

$$e = r_1 \, v' \tag{2.3}$$
$$d = r_2 e = r_2 r_1 v'$$
$$u = r_3 d = r_3 r_2 r_1 v'$$
$$c = r_4 u = r_4 r_3 r_2 r_1 v'$$
$$s = r_5 c = r_5 r_4 r_3 r_2 r_1 v'$$
$$t = r_6 s = r_6 r_5 r_4 r_3 r_2 r_1 v'$$
$$b = r_7 t = r_7 r_6 r_5 r_4 r_3 r_2 r_1 v'$$

This buildup is in slice factors of *mass* in a manner analogous to the buildup of coupling constants with spin degrees of freedom factors in section 4.2 of Blaha (2024f).

Following the arrows in Fig. 2.1 we see two patterns emerge. The set of angled arrows exhibit an increasing sequence of ratios. We form ratios of ratios and see approximate factors containing π appear:

$$
\begin{array}{cccc}
r_2 & r_4 & r_6 & \\
7.358 & 721.6 & 1819 & \\
& r_4/r_2 = 98.1 & r_6/r_4 = 2.52 & \\
& \cong 32\pi & \cong \pi &
\end{array}
\tag{2.4}
$$

The set of vertical arrows exhibit a decreasing sequence of ratios:

$$
\begin{array}{cccc}
r_1 & r_3 & r_5 & r_7 \\
5.81 & 0.468 & 0.0748 & 0.024 \\
& r_1/r_3 = 12.42 \quad r_3/r_5 = 6.28 & r_5/r_7 = 3.116 & \\
& \cong 4\pi \qquad\quad\; \cong 2\pi & \cong \pi &
\end{array}
\tag{2.5}
$$

The angled line data gives:

$$r_4 = 32\pi r_2 \qquad\qquad r_6 = \pi r_4 = 32\pi^2 r_2 \tag{2.6}$$

Note the powers of $\pi^{(i-2)/2}$ for $i = 4, 6$.
The vertical line data gives:

$$r_1 = 5.81$$

$$r_5 = \pi\, r_7 \qquad r_3 = 2\pi r_5 \qquad r_1 = 4\pi r_3 \qquad\qquad (2.7)$$

Using the above relations we find

$$r_1 = e/\,v' = .131/.02254 = 5.81 \cong 2\pi \qquad\qquad (2.8)$$
$$r_2 \cong 4/\pi\, r_1 \cong 8$$
$$r_3 \cong r_1/(4\pi) \cong 1/2$$
$$r_4 \cong 256\pi$$
$$r_5 \cong 1/(4\pi)$$
$$r_6 \cong 256\pi^2$$
$$r_7 \cong 1/(4\pi^2)$$

with $v' = 8.8 \times 10^{-5}$ GeV/c^2. Fig. 2.2 shows the resulting approximate mass values.

INITIAL ESTIMATE	**DEVIATION RATIO**	**DATA**
$e = r_1 v' = 2\pi v' = .553 \times 10^{-3}$	1.08	0.511×10^{-3} GeV/c^2
$d = r_2 e - r_2 r_1 v' = 2^4\pi v' = 4.42 \times 10^{-3}$	1.18	3.76×10^{-3} GeV/c^2
$u = r_3 d = r_3 r_2 r_1 v' = 2^3\pi v' = 2.51 \times 10^{-3}$	1.42	1.76×10^{-3} GeV/c^2
$c = r_4 u = r_4 r_3 r_2 r_1 v' = 2^{11}\pi^2 v' = 1.78$	1.4	1.27 GeV/c^2
$s = r_5 c = r_5 r_4 r_3 r_2 r_1 v' = 2^9\pi v' = 141.55 \times 10^{-3}$	1.49	95×10^{-3} GeV/c^2
$t = r_6 s = r_6 r_5 r_4 r_3 r_2 r_1 v' = 2^{17}\pi^3 v' = 357.64$	2.07	172.76 GeV/c^2
$b = r_7 t = r_7 r_6 r_5 r_4 r_3 r_2 r_1 v' = 2^{15}\pi v' = 9.06$	2.16	4.18 GeV/c^2

Figure 2.2. Approximate estimates of the fermion current masses. The estimates can be viewed as approximations to the experimental data based on powers of 2 and π.

The agreement of the estimates[12] formed from examining the data and the actual data is good for e and d; is less good for u, c, and s; and much less good for t and b. We remedy these anomalies in chapter 3.

---

[12] The limitation to powers of 2 and π times v' makes the estimates non-trivial.

3. Revised Estimate of Current Fermion Masses based on SU(2)/SU(3) Variation

In the preceding chapter we developed the forms of eight known fundamental fermion masses based on a constant times a power of 2 and a power of π. These forms are analogous to the forms of coupling constants that we recently found based on powers of 2 and e – the base of the natural logarithms. Together this harmony of mass and coupling constant forms lends indirect support to Cosmos Theory. It is somewhat remarkable that coupling constants are based on e and masses are based on π. They are not jointly based on both. It suggests there is a deeper, new layer of particle interactions which has yet to be elucidated.

We now turn to the possibility of improving the mass estimates of chapter 2 by the introduction of a factor giving a sliding scale of mass dependence in the ratio of an ElectroWeak SU(2) factor and a Strong Interaction SU(3) factor. We assume they contribute with equal strength to the e and d masses; with a mild dominance of ElectroWeak over Strong for u, c, and s quarks; and with dominance of Strong over ElectroWeak by a factor of 2 for t and b quarks.

We denote the ratio of the interactions with

$$\alpha/\beta \equiv \alpha_{SU(2)}(m)/\beta_{SU(3)}(m) \qquad (3.1)$$

where m is the mass of the relevant fermion. We note that the ratio of 2 for t and b reflects the ratio of the coupling constants ($1.21/0.63 = 1.92 \cong 2$) in Fig. 1.1. The value of α/β nicely progresses from 1 to ¾ to ½ as the fermion mass increases.

α/β	REVISED ESTIMATE	DEVIATION RATIO	DATA
	$v' \qquad\qquad\qquad = 8.8 \times 10^{-5}$ GeV/c^2		
1	$e = r_1 v' = 2\,\pi\,v' = .553 \times 10^{-3}$	1.08	0.511×10^{-3} GeV/c^2
1	$d = r_2 e = r_2 r_1 v' = 16\pi\,v' = 4.42 \times 10^{-3}$	1.18	3.76×10^{-3} GeV/c^2
¾	$u = r_3 d = r_3 r_2 r_1 v' = \text{¾} \times 2^3 \pi\,v' = 1.88 \times 10^{-3}$	1.06	1.76×10^{-3} GeV/c^2
¾	$c = r_4 u = r_4 r_3 r_2 r_1 v' = \text{¾} \times 2^{11} \pi^2\,v' = 1.335$	1.05	1.27 GeV/c^2
¾	$s = r_5 c = r_5 r_4 r_3 r_2 r_1 v' = \text{¾} \times 2^9 \pi\,v' = 106 \times 10^{-3}$	1.12	95×10^{-3} GeV/c^2
½	$t = r_6 s = r_6 r_5 r_4 r_3 r_2 r_1 v' = 2^{16} \pi^3\,v' = 178.8$	1.03	172.76 GeV/c^2
½	$b = r_7 t = r_7 r_6 r_5 r_4 r_3 r_2 r_1 v' = 2^{14} \pi\,v' = 4.53$	1.08	4.18 GeV/c^2

Figure 3.1. Revised fermion masses based on α/β.

Strong SU(3) dominates at high mass; ElectroWeak SU(2) dominates at low mass. All estimates in Fig. 3.1 are close to the experimental values. Thus we have a representation of fermion masses in terms of powers of 2, powers of π, and the value of v' - comparable to coupling constants where coupling constants are powers of 2 multiplied by e – the base of natural logarithms. This suggests that Cosmos Theory is at the base of both coupling constant and fermion mass values. Why one uses π and the other uses e - the base of natural logarithms, remains to be understood.

4. Vector Boson Masses

The calculation of fermion mass data as functions of powers of 2 and π may be extended to the case of the vector boson masses in ElectroWeak theory. We previously showed the Weinberg angle θ_W, which is defined by

$$e^2/g^2 = \sin^2 \theta_W \qquad (4.1)$$

using the coupling constants of Fig. 1.1, sets

$$\sin \theta_W = 0.5 \qquad (4.2)$$

in close approximation to the known value of θ_W (where $\sin^2 \theta_W = 0.231$).
      The W and Z masses are related by

$$m_W^2/m_Z^2 = \cos^2 \theta_W = 0.75 \qquad (4.3)$$

They are known to have the values

$$W = 80.377 \qquad (4.4)$$
$$Z = 91.19$$

where we use their symbols to represent their masses.
      Noting that the t quark mass numerically satisfies

$$t/Z = 178.8/91.19 = 1.96 \cong 2 \qquad (4.5)$$

we can use the Fig. 3.1 value of t to write

$$Z = t/2 = 2^{15}\pi^3 \, v' \qquad (4.6)$$

thus giving an approximate expression for Z in terms of powers of 2 and π. Eq. 4.4 implies the W mass

$$W^2 = 0.75Z^2 = \tfrac{3}{4} \, 2^{30}\pi^6 \, v'^2 \qquad (4.7)$$

or

$$W = \sqrt{\tfrac{3}{4}} \, 2^{15}\pi^6 \, v' \qquad (4.8)$$

Thus the W and Z masses have simple representations in terms of powers of 2 and π analogous to those of the fermion masses.

5. Four Generations of Fermions

We have suggested that the Cosmos Theory portrayal of generations of fermions (Fig. 1.3) should be preferred especially in view of the patterns of first generation current fermion masses. (Fig. 3.1)

We now consider the extension of the first generation mass patterns and spectrum to the second, third and fourth generations. The masses of the μ and τ leptons (of the second and third generations) are known. We can use their mass ratios to extend the quark and neutrino masses to the second and third generations.

The μ and τ mass ratios can be approximated as

$$\mu/e = 207 \cong 2\pi^4 = 195 \qquad \text{with a 1.06 factor deviation}$$
$$\tau/\mu = 16.76 \cong 2^4 = 16 \qquad \text{with a 1.05 factor deviation}$$

Based on these ratios we take the estimates in Figs. 1.4 and 3.1. and determine the second and third generation fermion masses.

SECOND GENERATION

α/β

$$\nu'_2 = 0.0172 \text{ GeV/c}^2$$
$$\nu_2 = 7.56 \times 10^{-9} \text{ GeV/c}^2 = 7.56 \text{ ev/c}^2$$

1 $\qquad e_2 = 2^2\pi^5 \nu' = 0.108 \text{ GeV/c}^2$

1 $\qquad d_2 = 32\pi^5 \nu' = 0.86 \text{ GeV/c}^2$

¾ $\qquad u_2 = \frac{3}{4} \times 2^5\pi^5 \nu' = 0.367 \text{ GeV/c}^2$

¾ $\qquad c_2 = \frac{3}{4} \times 2^{12}\pi^6 \nu' = 260 \text{ GeV/c}^2$

¾ $\qquad s_2 = \frac{3}{4} \times 2^{10} \pi^5 \nu' = 20.6 \text{ GeV/c}^2$

½ $\qquad t_2 = 2^{17}\pi^7 \nu' = 34{,}866 \text{ GeV/c}^2$

½ $\qquad b_2 = 2^{15} \pi^5 \nu' = 883 \text{ GeV/c}^2$

Figure 5.1. Projected Second Generation fermion masses using $\nu' = 8.8 \times 10^{-5}$ GeV/c^2.

THIRD GENERATION

α/β

$$v'_3 = 0.275 \text{ GeV/c}^2$$

$v_3 = 1.21 \times 10^{-7} \text{ GeV/c}^2 = 121 \text{ ev/c}^2$ This value places the total of neutrino masses above the 50 eV/c^2 limit which might have resulted in the collapse of the universe unless the neutrino is unstable.

1	$e_3 = 2^2\pi^5 v' = 1.73 \text{ GeV/c}^2$
1	$d_3 = 32\pi^5 v' = 13.76 \text{ GeV/c}^2$
¾	$u_3 = \text{¾} \times 2^5\pi^5 v' = 5.87 \text{ GeV/c}^2$
¾	$c_3 = \text{¾} \times 2^{12}\pi^6 v' = 4160 \text{ GeV/c}^2$
¾	$s_3 = \text{¾} \times 2^{10} \pi^5 v' = 329.6 \text{ GeV/c}^2$
½	$t_3 = 2^{17}\pi^7 v' = 557{,}856 \text{ GeV/c}^2$
½	$b_3 = 2^{15} \pi^5 v' = 14{,}128 \text{ GeV/c}^2$

Figure 5.2. Projected Third Generation fermion masses using $v' = 8.8 \times 10^{-5} \text{ GeV/c}^2$.

REFERENCES

Akhiezer, N. I., Frink, A. H. (tr), 1962, *The Calculus of Variations* (Blaisdell Publishing, New York, 1962).

Bjorken, J. D., Drell, S. D., 1964, *Relativistic Quantum Mechanics* (McGraw-Hill, New York, 1965).

Bjorken, J. D., Drell, S. D., 1965, *Relativistic Quantum Fields* (McGraw-Hill, New York, 1965).

Blaha, S., 1995, *C++ for Professional Programming* (International Thomson Publishing, Boston, 1995).

_______, 1998, *Cosmos and Consciousness* (Pingree-Hill Publishing, Auburn, NH, 1998 and 2002).

_______, 2002, *A Finite Unified Quantum Field Theory of the Elementary Particle Standard Model and Quantum Gravity Based on New Quantum Dimensions™ & a New Paradigm in the Calculus of Variations* (Pingree-Hill Publishing, Auburn, NH, 2002).

_______, 2004, *Quantum Big Bang Cosmology: Complex Space-time General Relativity, Quantum Coordinates™ Dodecahedral Universe, Inflation, and New Spin 0, ½, 1 & 2 Tachyons & Imagyons* (Pingree-Hill Publishing, Auburn, NH, 2004).

_______, 2005a, *Quantum Theory of the Third Kind: A New Type of Divergence-free Quantum Field Theory Supporting a Unified Standard Model of Elementary Particles and Quantum Gravity based on a New Method in the Calculus of Variations* (Pingree-Hill Publishing, Auburn, NH, 2005).

_______, 2005b, *The Metatheory of Physics Theories, and the Theory of Everything as a Quantum Computer Language* (Pingree-Hill Publishing, Auburn, NH, 2005).

_______, 2005c, *The Equivalence of Elementary Particle Theories and Computer Languages: Quantum Computers, Turing Machines, Standard Model, Superstring Theory, and a Proof that Gödel's Theorem Implies Nature Must Be Quantum* (Pingree-Hill Publishing, Auburn, NH, 2005).

_______, 2006a, *The Foundation of the Forces of Nature* (Pingree-Hill Publishing, Auburn, NH, 2006).

_______, 2006b, *A Derivation of ElectroWeak Theory based on an Extension of Special Relativity; Black Hole Tachyons; & Tachyons of Any Spin.* (Pingree-Hill Publishing, Auburn, NH, 2006).

_______, 2007a, *Physics Beyond the Light Barrier: The Source of Parity Violation, Tachyons, and A Derivation of Standard Model Features* (Pingree-Hill Publishing, Auburn, NH, 2007).

_______, 2007b, *The Origin of the Standard Model: The Genesis of Four Quark and Lepton Species, Parity Violation, the ElectroWeak Sector, Color SU(3), Three Visible Generations of Fermions, and One Generation of Dark Matter with Dark Energy* (Pingree-Hill Publishing, Auburn, NH, 2007).

_______, 2008a, *A Direct Derivation of the Form of the Standard Model From GL(16) (Pingree-Hill Publishing, Auburn, NH, 2008).*

_______, 2008b, *A Complete Derivation of the Form of the Standard Model With a New Method to Generate Particle Masses Second Edition* (Pingree-Hill Publishing, Auburn, NH, 2008)

_______, 2009, *The Algebra of Thought & Reality: The Mathematical Basis for Plato's Theory of Ideas, and Reality Extended to Include A Priori Observers and Space-Time Second Edition* (Pingree-Hill Publishing, Auburn, NH, 2009).

_______, 2010a, *Operator Metaphysics: A New Metaphysics Based on a New Operator Logic and a New Quantum Operator Logic that Lead to a Mathematical Basis for Plato's Theory of Ideas and Reality* (Pingree-Hill Publishing, Auburn, NH, 2010).

______, 2010b, *The Standard Model's Form Derived from Operator Logic, Superluminal Transformations and GL(16)* (Pingree-Hill Publishing, Auburn, NH, 2010).

______, 2010c, *SuperCivilizations: Civilizations as Superorganisms* (McMann-Fisher Publishing, Auburn, NH, 2010).

______, 2011a, *21ˢᵗ Century Natural Philosophy Of Ultimate Physical Reality* (McMann-Fisher Publishing, Auburn, NH, 2011).

______, 2011b, *All the Universe! Faster Than Light Tachyon Quark Starships & Particle Accelerators with the LHC as a Prototype Starship Drive Scientific Edition* (Pingree-Hill Publishing, Auburn, NH, 2011).

______, 2011c, *From Asynchronous Logic to The Standard Model to Superflight to the Stars* (Blaha Research, Auburn, NH, 2011).

______, 2012a, *From Asynchronous Logic to The Standard Model to Superflight to the Stars volume 2: Superluminal CP and CPT, U(4) Complex General Relativity and The Standard Model, Complex Vierbein General Relativity, Kinetic Theory, Thermodynamics* (Blaha Research, Auburn, NH, 2012).

______, 2012b, *Standard Model Symmetries, And Four And Sixteen Dimension Complex Relativity; The Origin Of Higgs Mass Terms* (Blaha Reasearch, Auburn, NH, 2012).

______, 2013a, *Multi-Stage Space Guns, Micro-Pulse Nuclear Rockets, and Faster-Than-Light Quark-Gluon Ion Drive Starships* (Blaha Research, Auburn, NH, 2013).

______, 2013b, *The Bridge to Dark Matter; A New Sibling Universe; Dark Energy; Inflatons; Quantum Big Bang; Superluminal Physics; An Extended Standard Model Based on Geometry* (Blaha Reasearch, Auburn, NH, 2013).

______, 2014a, *Universes and Megaverses: From a New Standard Model to a Physical Megaverse; The Big Bang; Our Sibling Universe's Wormhole; Origin of the Cosmological Constant, Spatial Asymmetry of the Universe, and its Web of Galaxies; A Baryonic Field between Universes and Particles; Megaverse Extended Wheeler-DeWitt Equation* (Blaha Reasearch, Auburn, NH, 2014).

______, 2014b, *All the Megaverse! Starships Exploring the Endless Universes of the Cosmos Using the Baryonic Force* (Blaha Research, Auburn, NH, 2014).

______, 2014c, *All the Megaverse! II Between Megaverse Universes: Quantum Entanglement Explained by the Megaverse Coherent Baryonic Radiation Devices – PHASERs Neutron Star Megaverse Slingshot Dynamics Spiritual and UFO Events, and the Megaverse Microscopic Entry into the Megaverse* (Blaha Research, Auburn, NH, 2014).

______, 2015a, *PHYSICS IS LOGIC PAINTED ON THE VOID: Origin of Bare Masses and The Standard Model in Logic, U(4) Origin of the Generations, Normal and Dark Baryonic Forces, Dark Matter, Dark Energy, The Big Bang, Complex General Relativity, A Megaverse of Universe Particles* (Blaha Research, Auburn, NH, 2015).

______, 2015b, *PHYSICS IS LOGIC Part II: The Theory of Everything, The Megaverse Theory of Everything, U(4)⊗U(4) Grand Unified Theory (GUT), Inertial Mass = Gravitational Mass, Unified Extended Standard Model and a New Complex General Relativity with Higgs Particles, Generation Group Higgs Particles* (Blaha Research, Auburn, NH, 2015).

______, 2015c, *The Origin of Higgs ("God") Particles and the Higgs Mechanism: Physics is Logic III, Beyond Higgs – A Revamped Theory With a Local Arrow of Time, The Theory of Everything Enhanced, Why Inertial Frames are Special, Universes of the Mind* (Blaha Research, Auburn, NH, 2015).

______, 2015d, *The Origin of the Eight Coupling Constants of The Theory of Everything: U(8) Grand Unified Theory of Everything (GUTE), S^8 Coupling Constant Symmetry, Space-Time Dependent Coupling Constants, Big Bang Vacuum Coupling Constants, Physics is Logic IV* (Blaha Research, Auburn, NH, 2015).

______, 2016a, *New Types of Dark Matter, Big Bang Equipartition, and A New U(4) Symmetry in the Theory of Everything: Equipartition Principle for Fermions, Matter is 83.33% Dark, Penetrating the Veil of the Big Bang, Explicit QFT Quark Confinement and Charmonium, Physics is Logic V* (Blaha Research, Auburn, NH, 2016).

______, 2016b, *The Periodic Table of the 192 Quarks and Leptons in The Theory of Everything: The U(4) Layer Group, Physics is Logic VI* (Blaha Research, Auburn, NH, 2016).

______, 2016c, *New Boson Quantum Field Theory, Dark Matter Dynamics, Dark Matter Fermion Layer Mixing, Genesis of Higgs Particles, New Layer Higgs Masses, Higgs Coupling Constants, Non-Abelian Higgs Gauge Fields, Physics is Logic VII* (Blaha Research, Auburn, NH, 2016).

______, 2016d, *Unification of the Strong Interactions and Gravitation: Quark Confinement Linked to Modified Short-Distance Gravity; Physics is Logic VIII* (Blaha Research, Auburn, NH, 2016).

______, 2016e, *MoND: Unification of the Strong Interactions and Gravitation II, Quark Confinement Linked to Large-Scale Gravity, Physics is Logic IX* (Blaha Research, Auburn, NH, 2016).

______, 2016f, *CQ Mechanics: A Unification of Quantum & Classical Mechanics, Quantum/Semi-Classical Entanglement, Quantum/Classical Path Integrals, Quantum/Classical Chaos* (Blaha Research, Auburn, NH, 2016).

______, 2016g, *GEMS Unified Gravity, ElectroMagnetic and Strong Interactions: Manifest Quark Confinement, A Solution for the Proton Spin Puzzle, Modified Gravity on the Galactic Scale* (Pingree Hill Publishing, Auburn, NH, 2016).

______, 2016h, *Unification of the Seven Boson Interactions based on the Riemann-Christoffel Curvature Tensor* (Pingree Hill Publishing, Auburn, NH, 2016).

______, 2017a, *Unification of the Eleven Boson Interactions based on 'Rotations of Interactions'* (Pingree Hill Publishing, Auburn, NH, 2017).

______, 2017b, *The Origin of Fermions and Bosons, and Their Unification* (Pingree Hill Publishing, Auburn, NH, 2017).

______, 2017c, *Megaverse: The Universe of Universes* (Pingree Hill Publishing, Auburn, NH, 2017).

______, 2017d, *SuperSymmetry and the Unified SuperStandard Model* (Pingree Hill Publishing, Auburn, NH, 2017).

______, 2017e, *From Qubits to the Unified SuperStandard Model with Embedded SuperStrings: A Derivation* (Pingree Hill Publishing, Auburn, NH, 2017).

______, 2017f, *The Unified SuperStandard Model in Our Universe and the Megaverse: Quarks, … ,* (Pingree Hill Publishing, Auburn, NH, 2017).

______, 2018a, *The Unified SuperStandard Model and the Megaverse SECOND EDITION A Deeper Theory based on a New Particle Functional Space that Explicates Quantum Entanglement Spookiness (Volume 1)* (Pingree Hill Publishing, Auburn, NH, 2018).

______, 2018b, *Cosmos Creation: The Unified SuperStandard Model, Volume 2, SECOND EDITION* (Pingree Hill Publishing, Auburn, NH, 2018).

______, 2018c, *God Theory* (Pingree Hill Publishing, Auburn, NH, 2018).

______, 2018d, *Immortal Eye: God Theory: Second Edition* (Pingree Hill Publishing, Auburn, NH, 2018).

______, 2018e, *Unification of God Theory and Unified SuperStandard Model THIRD EDITION* (Pingree Hill Publishing, Auburn, NH, 2018).

______, 2019a, *Calculation of: QED α = 1/137, and Other Coupling Constants of the Unified SuperStandard Theory* (Pingree Hill Publishing, Auburn, NH, 2019).

______, 2019b, *Coupling Constants of the Unified SuperStandard Theory SECOND EDITION* (Pingree Hill Publishing, Auburn, NH, 2019).

______, 2019c, *New Hybrid Quantum Big_Bang–Megaverse_Driven Universe with a Finite Big Bang and an Increasing Hubble Constant* (Pingree Hill Publishing, Auburn, NH, 2019).

______, 2019d, *The Universe, The Electron and The Vacuum* (Pingree Hill Publishing, Auburn, NH, 2019).

______, 2019e, *Quantum Big Bang – Quantum Vacuum Universes (Particles)* (Pingree Hill Publishing, Auburn, NH, 2019).

______, 2019f, *The Exact QED Calculation of the Fine Structure Constant Implies ALL 4D Universes have the Same Physics/Life Prospects* (Pingree Hill Publishing, Auburn, NH, 2019).

______, 2019g, *Unified SuperStandard Theory and the SuperUniverse Model: The Foundation of Science* (Pingree Hill Publishing, Auburn, NH, 2019).

______, 2020a, *Quaternion Unified SuperStandard Theory (The QUeST) and Megaverse Octonion SuperStandard Theory (MOST)* (Pingree Hill Publishing, Auburn, NH, 2020).

______, 2020b, *United Universes Quaternion Universe - Octonion Megaverse* (Pingree Hill Publishing, Auburn, NH, 2020).

______, 2020c, *Unified SuperStandard Theories for Quaternion Universes & The Octonion Megaverse* (Pingree Hill Publishing, Auburn, NH, 2020).

______, 2020d, *The Essence of Eternity: Quaternion & Octonion SuperStandard Theories* (Pingree Hill Publishing, Auburn, NH, 2020).

______, 2020e, *The Essence of Eternity II* (Pingree Hill Publishing, Auburn, NH, 2020).

______, 2020f, *A Very Conscious Universe* (Pingree Hill Publishing, Auburn, NH, 2020).

______, 2020g, *Hypercomplex Universe* (Pingree Hill Publishing, Auburn, NH, 2020).

______, 2020h, *Beneath the Quaternion Universe* (Pingree Hill Publishing, Auburn, NH, 2020).

______, 2020i, *Why is the Universe Real? From Quaternion & Octonion to Real Coordinates* (Pingree Hill Publishing, Auburn, NH, 2020).

______, 2020j, *The Origin of Universes: of Quaternion Unified SuperStandard Theory (QUeST); and of the Octonion Megaverse (UTMOST)* (Pingree Hill Publishing, Auburn, NH, 2020).

______, 2020k, *The Seven Spaces of Creation: Octonion Cosmology* (Pingree Hill Publishing, Auburn, NH, 2020).
______, 2020l, *From Octonion Cosmology to the Unified SuperStandard Theory of Particles* (Pingree Hill Publishing, Auburn, NH, 2020).

______, 2021a, *Pioneering the Cosmos* (Pingree Hill Publishing, Auburn, NH, 2021).

______, 2021b, *Pioneering the Cosmos II* (Pingree Hill Publishing, Auburn, NH, 2021).

______, 2021c, *Beyond Octonion Cosmology* (Pingree Hill Publishing, Auburn, NH, 2021).

______, 2021d, *Universes are Particles* (Pingree Hill Publishing, Auburn, NH, 2021).

______, 2021e, *Octonion-like dna-based life, Universe expansion is decay, Emerging New Physics* (Pingree Hill Publishing, Auburn, NH, 2021).

______, 2021f, *The Science of Creation New Quantum Field Theory of Spaces* (Pingree Hill Publishing, Auburn, NH, 2021).

______, 2021g, *Quantum Space Theory With Application to Octonion Cosmology & Possibly To Fermionic Condensed Matter* (Pingree Hill Publishing, Auburn, NH, 2021).

______, 2021h, *21st Century Natural Philosophy of Octonion Cosmology , and Predestination, Fate, and Free Will* (Pingree Hill Publishing, Auburn, NH, 2021).

______, 2021i, *Beyond Octonion Cosmology II : Origin of the Quantum; A New Generalized Field Theory (GiFT); A Proof of the Spectrum of Universes; Atoms in Higher Universes* (Pingree Hill Publishing, Auburn, NH, 2021).

______, 2021j, *Integration of General Relativity and Quantum Theory: Octonion Cosmology, GiFT, Creation/Annihilation Spaces CASe, Reduction of Spaces to a Few Fermions and Symmetries in Fundamental Frames* (Pingree Hill Publishing, Auburn, NH, 2021).

______, 2022a, *New View of Octonion Cosmology Based on the Unification of General Relativity and Quantum Theory* (Pingree Hill Publishing, Auburn, NH, 2022).

______, 2022b, *The Dust Beneath Hypercomplex Cosmology* (Pingree Hill Publishing, Auburn, NH, 2022).

______, 2022c, *Passing Through Nature to Eternity: ProtoCosmos, HyperCosmos, Unified SuperStandard Theory* (Pingree Hill Publishing, Auburn, NH, 2022).

______, 2022d, *HyperCosmos Fractionation and Fundamental Reference Frame Based Unification: Particle Inner Space Basis of Parton and Dual Resonance Models* (Pingree Hill Publishing, Auburn, NH, 2022).

______, 2022e, *A New UniDimension ProtoCosmos and SuperString F-Theory Relation to the HyperCosmos* (Pingree Hill Publishing, Auburn, NH, 2022).

______, 2022f, *The Cosmic Panorama: ProtoCosmos, HyperCosmos,Unified SuperStandard Theory (UST) Derivation* (Pingree Hill Publishing, Auburn, NH, 2022).

______, 2022g, *Ultimate Origin: ProtoCosmos and HyperCosmos* (Pingree Hill Publishing, Auburn, NH, 2022).

______, 2023a, *UltraUnification and the Generation of the Cosmos* (Pingree Hill Publishing, Auburn, NH, 2023).

______, 2023b, *God and and Cosmos Theory* (Pingree Hill Publishing, Auburn, NH, 2023).

______, 2023c, *A New Completely Geometric SU(8) Cosmos Theory; New PseudoFermion Fields; Fibonacci-like Dimension Arrays; Ramsey Number Approximation* (Pingree Hill Publishing, Auburn, NH, 2023).

______, 2023d, *Newton's Apple is Now the Fermion* (Pingree Hill Publishing, Auburn, NH, 2023).

______, 2023e,*Cosmos Theory: The Sub-Particle Gambol Model* (Pingree Hill Publishing, Auburn, NH, 2023).

______, 2024a, *Cosmos-Universe-Particle-Gambol Theory* (Pingree Hill Publishing, Auburn, NH, 2024).

______, 2024b, *Fractal Cosmos Theory* (Pingree Hill Publishing, Auburn, NH, 2024).

______, 2024c, *Fractal Cosmic Curve: Tensor-Based CosmosTheory* (Pingree Hill Publishing, Auburn, NH, 2024).

REFERENCES

_____, 2024d, *The Eternal Form of Cosmos Theory* (Pingree Hill Publishing, Auburn, NH, 2024).

_____, 2024e, *The Eternal Form of Cosmos Theory Third Edition* (Pingree Hill Publishing, Auburn, NH, 2024).

_____, 2024f, *Fundamental Constants of Cosmos Theory and The Standard Model* (Pingree Hill Publishing, Auburn, NH, 2024).

Eddington, A. S., 1952, *The Mathematical Theory of Relativity* (Cambridge University Press, Cambridge, U.K., 1952).

Fant, Karl M., 2005, *Logically Determined Design: Clockless System Design With NULL Convention Logic* (John Wiley and Sons, Hoboken, NJ, 2005).

Feinberg, G. and Shapiro, R., 1980, *Life Beyond Earth: The Intelligent Earthlings Guide to Life in the Universe* (William Morrow and Company, New York, 1980).

Gelfand, I. M., Fomin, S. V., Silverman, R. A. (tr), 2000, *Calculus of Variations* (Dover Publications, Mineola, NY, 2000).

Giaquinta, M., Modica, G., Souchek, J., 1998, *Cartesian Coordinates in the Calculus of Variations* Volumes I and II (Springer-Verlag, New York, 1998).

Giaquinta, M., Hildebrandt, S., 1996, *Calculus of Variations* Volumes I and II (Springer-Verlag, New York, 1996).

Gradshteyn, I. S. and Ryzhik, I. M., 1965, *Table of Integrals, Series, and Products* (Academic Press, New York, 1965).

Heitler, W., 1954, *The Quantum Theory of Radiation* (Claendon Press, Oxford, UK, 1954).

Huang, Kerson, 1992, *Quarks, Leptons & Gauge Fields 2nd Edition* (World Scientific Publishing Company, Singapore, 1992).

Jost, J., Li-Jost, X., 1998, *Calculus of Variations* (Cambridge University Press, New York, 1998).

Kaku, Michio, 1993, *Quantum Field Theory*, (Oxford University Press, New York, 1993).

Kirk, G. S. and Raven, J. E., 1962, *The Presocratic Philosophers* (Cambridge University Press, New York, 1962).

Landau, L. D. and Lifshitz, E. M., 1987, *Fluid Mechanics 2nd Edition*, (Pergamon Press, Elmsford, NY, 1987).

Rescher, N., 1967, *The Philosophy of Leibniz* (Prentice-Hall, Englewood Cliffs, NJ, 1967).

Riesz, Frigyes and Sz.-Nagy, Béla, 1990, *Functional Analysis* (Dover Publications, New York, 1990).

Sakurai, J. J., 1964, *Invariance Principles and Elementary Particles* (Princeton University Press, Princeton, NJ, 1964).

Weinberg, S., 1972, *Gravitation and Cosmology* (John Wiley and Sons, New York, 1972).

Weinberg, S., 1995, *The Quantum Theory of Fields Volume I* (Cambridge University Press, New York, 1995).

INDEX

About the Author

Stephen Blaha is a well-known Physicist and Man of Letters with interests in Science, Society and civilization, the Arts, and Technology. He had an Alfred P. Sloan Foundation scholarship in college. He received his Ph.D. in Physics from Rockefeller University. He has served on the faculties of several major universities. He was also a Member of the Technical Staff at Bell Laboratories, a manager at the Boston Globe Newspaper, a Director at Wang Laboratories, and President of Blaha Software Inc. and of Janus Associates Inc. (NH).

Among other achievements he was a co-discoverer of the "r potential" for heavy quark binding developing the first (and still the only demonstrable) non-Aeolian gauge theory with an "r" potential; first suggested the existence of topological structures in superfluid He-3; first proposed Yang-Mills theories would appear in condensed matter phenomena with non-scalar order parameters; first developed a grammar-based formalism for quantum computers and applied it to elementary particle theories; first developed a new form of quantum field theory without divergences (thus solving a major 60 year old problem that enabled a unified theory of the Standard Model and Quantum Gravity without divergences to be developed); first developed a formulation of complex General Relativity based on analytic continuation from real space-time; first developed a generalized non-homogeneous Robertson-Walker metric that enabled a quantum theory of the Big Bang to be developed without singularities at t = 0; first generalized Cauchy's theorem and Gauss' theorem to complex, curved multi-dimensional spaces; received Honorable Mention in the Gravity Research Foundation Essay Competition in 1978; first developed a physically acceptable theory of faster-than-light particles; first derived a composition of extremums method in the Calculus of Variations; first quantitatively suggested that inflationary periods in the history of the universe were not needed; first proved Gödel's Theorem implies Nature must be quantum; provided a new alternative to the Higgs Mechanism, and Higgs particles, to generate masses; first showed how to resolve logical paradoxes including Gödel's Undecidability Theorem by developing Operator Logic and Quantum Operator Logic; first developed a quantitative harmonic oscillator-like model of the life cycle, and interactions, of civilizations; first showed how equations describing superorganisms also apply to civilizations. A recent book shows his theory applies successfully to the past 14 years of history and to *new* archaeological data on Andean and Mayan civilizations as well as Early Anatolian and Egyptian civilizations.

He first developed an axiomatic derivation of the form of The Standard Model from geometry – space-time properties – The Unified SuperStandard Model. It unifies all the known forces of Nature. It also has a Dark Matter sector that includes a Dark ElectroWeak sector with Dark doublets and Dark gauge interactions. It uses quantum coordinates to remove infinities that crop up in most interacting quantum field theories and additionally to remove the infinities that appear in the Big Bang and generate inflationary growth of the universe. It shows gravity has a MOND-like form without sacrificing Newton's Laws. It relates the interactions of the MOND-like sector of gravity with the r-potential of Quark Confinement. The axioms of the theory lead to the question of their origin. We suggest in the preceding edition of this book it can be attributed to an entity with God-like properties. We explore these properties in "God Theory" and show they predict that the Cosmos exists forever although individual universes (or incarnations of our universe) "come and go." Several other important results emerge from God Theory such a functionally triune God. The Unified SuperStandard Theory has many other important parts described in the Current Edition of *The Unified SuperStandard Theory* and expanded in subsequent volumes.

Blaha has had a major impact on a succession of elementary particle theories: his Ph.D. thesis (1970), and papers, showed that quantum field theory calculations to all orders in ladder approximations could not give scaling deep inelastic electron-nucleon scattering. He later showed the eigenvalue equation for the fine structure constant α in Johnson-Baker-Willey QED had a zero at $\alpha = 1$ not 1/137 by solving the Schwinger-Dyson equations to all orders in an approximation that agreed with exact results to 4th order in α thus ending interest in this theory. In 1979 at Prof. Ken Johnson's (MIT) suggestion he calculated the proton-neutron mass difference in the MIT bag model and found the result had the wrong sign reducing interest in the bag model. These results all appear in Physical Review papers. In the 2000's he repeatedly pointed out the shortcomings of SuperString theory and showed that The Standard Model's form could be derived from space-time geometry by an extension of Lorentz transformations to faster than light transformations. This deeper space-time basis greatly increases the possibility that it is part of THE fundamental theory. Recently, Blaha showed that the Weak interactions differed significantly from the Strong, electromagnetic and gravitation interactions in important respects while these interactions had similar features, and suggested that ElectroWeak theory, which is essentially a glued union of the Weak interactions and Electromagnetism, possibly modulo unknown Higgs particle features, be replaced by a unified theory of the other interactions combined with a stand-alone Weak interaction theory. Blaha also showed that, if Charmonium calculations are taken seriously, the Strong interaction coupling constant is only a factor of five larger than the electromagnetic coupling constant, and thus Strong interaction perturbation theory would make sense and yield physically meaningful results.

In graduate school (1965-71) he wrote substantial papers in elementary particles and group theory: The Inelastic E- P Structure Functions in a Gluon Model. Phys. Lett. B40:501-502,1972; Deep-Inelastic E-P Structure Functions In A Ladder Model With Spin 1/2 Nucleons, Phys.Rev. D3:510-523,1971; Continuum Contributions To The Pion Radius, Phys. Rev. 178:2167-2169,1969; Character Analysis of U(N) and SU(N), J. Math. Phys. <u>10</u>, 2156 (1969); and The Calculation of the Irreducible Characters of the Symmetric Group in Terms of the Compound Characters, (Published as Blaha's Lemma in D. E. Knuth's book: *The Art of Computer Programming Vols. 1 – 4*).
In the early 1980's Blaha was also a pioneer in the development of UNIX for financial, scientific and Internet applications: benchmarked UNIX versions showing that block size was critical for UNIX performance, developing financial modeling software,

starting database benchmarking comparison studies, developing Internet-like UNIX networking (1982) and developing a hybrid shell programming technique (1982) that was a precursor to the PERL programming language. He was also the manager of the AT&T ten-year future products development database. His work helped lead to commercial UNIX on computers such as Sun Micros, IBM AIX minis, and Apple computers.

In the 1980's he pioneered the development of PC Desktop Publishing on laser printers and was nominated for three "Awards for Technical Excellence" in 1987 by PC Magazine for PC software products that he designed and developed.

Recently he has developed a theory of Megaverses – actual universes of which our universe is one – with quantum particle-like properties based on the Wheeler-DeWitt equation of Quantum Gravity. He has developed a theory of a baryonic force, which had been conjectured many years ago, and estimated the strength of the force based on discrepancies in measurements of the gravitational constant G. This force, operative in D-dimensional space, can be used to escape from our universe in "uniships" which are the equivalent of the faster-than-light starships proposed in the author's earlier books. Thus travel to other universes, as well as to other stars is possible.

Blaha also considered the complexified Wheeler-DeWitt equation and showed that its limitation to real-valued coordinates and metrics generated a Cosmological Constant in the Einstein equations.

The author has also recently written a series of books on the serious problems of the United States and their solution as well as a book on the decline of Mankind that will follow from current social and genetic trends in Mankind.

In the past twenty years Dr. Blaha has written over 80 books on a wide range of topics. Some recent major works are: *From Asynchronous Logic to The Standard Model to Superflight to the Stars, All the Universe!, SuperCivilizations: Civilizations as Superorganisms, America's Future: an Islamic Surge, ISIS, al Qaeda, World Epidemics, Ukraine, Russia-China Pact, US Leadership Crisis, The Rises and Falls of Man – Destiny – 3000 AD: New Support for a Superorganism MACRO-THEORY of CIVILIZATIONS From CURRENT WORLD TRENDS and NEW Peruvian, Pre-Mayan, Mayan, Anatolian, and Early Egyptian Data, with a Projection to 3000 AD,* and *Mankind in Decline: Genetic Disasters, Human-Animal Hybrids, Overpopulation, Pollution, Global Warming, Food and Water Shortages, Desertification, Poverty, Rising Violence, Genocide, Epidemics, Wars, Leadership Failure.*

He has taught approximately 4,000 students in undergraduate, graduate, and postgraduate corporate education courses primarily in major universities, and large companies and government agencies.

He developed a quantum theory, The Unified SuperStandard Theory (UST), which describes elementary particles in detail without the difficulties of conventional quantum field theory. He found that the internal symmetries of this theory could be exactly derived from an octonion theory called QUeST. He further found that another octonion theory (UTMOST) describes the Megaverse. It can hold QUeST universes such as our own universe. It has an internal symmetry structure which is a superset of the QUeST internal symmetries.

Recently he developed Octonion Cosmology. He replaced it with HyperCosmos theory, which has significantly better features. He developed a fractionalization process for dimensions, particles and symmetry groups. He also described transformation that reduced particles and dimensions to a far more compact form. He also developed a precursor theory ProtoCosmos that leads to the HyperCosmos.

The author showed that space-time and Internal Symmetries can be unified in any of the ten HyperCosmos spaces in their associated HyperUnification spaces. The combined set of HyperUnification spaces enable all HyperCosmos dimensions to be obtained by a General Relativistic transformation from one primordial dimension in the 42 space-time dimension unified HyperUnification space.

At present the author devel;oped the Cosmos Theory that incorporates ProtoCosmos Theory, HyperCosmos Theory, Limos Theory, Second Kind HyperCosmos Theory and HyperUnification Spaces. He has introduced PseudoFermion wave functions and theory, He has related Cosmos Theory to Regge trajectories of spaces, parton theory, Veneziano amplitudes, Fibonacci numbers and Ramsey numbers. He has calculated an approximation to the difficult R(n,n) Ramsey numbers.

He has developed a Gambol Model that successfully accounts for e-p deep inelastic scattering, fundamental particle resonances, hadron scattering, and the inner structure of particles based on confinement through Casimir forces of ideal gambol gases. The Gambol Planckian Distribution was derived.

He has applied the Gambol Model to particles, universes, and the Cosmos of universes. He showed that the Cosmos may have a distribution of 23 universes corresponding to various Cosmos spaces.

Recently he showed that Cosmos Theory follows from the number of independent asymmetric tensors in a dimension r. He also showed the close parallel between the form of γ-matrices and Cosmos Theory dimension arrays. The closeness suggested that dimension arrays have the same importance as γ-matrices for fermions.

He demonstrated that the pressure of fermions within a space of dimension r balances the Casimir vacuum energy force for 18 dimensions. He showed that $2e\pi = 17.02$ marks the critical point where pressure balances Casimir force, which implies $r = 18$ is the highest dimension Physical Cosmos space. The dimension $2e\pi$ appears to set the approximate dimension for Cosmos spaces with dimension array size $2^{r+4} \cong (17.02/8)^{r+4} \cong (e\pi/4)^{r+4} \cong 2.13^{r+4}$.

Now he has found the sequence of Coupling Constant values in the Standard Model and UST.